think green

Water Conservation

SADDLEBACK
EDUCATIONAL PUBLISHING
Three Watson
Irvine, CA 92618-2767
Website: www.sdlback.com

ISBN-13: 978-1-59905-458-2
ISBN-10: 1-59905-458-2
eBook: 978-1-60291-784-2

Printed in China
15 14 13 12 11 1 2 3 4 5

Contents

The Earth's Water

Water is needed for all life on Earth. Almost 71% of Earth's surface is covered with water. Water is found in many forms. These include groundwater, surface water, and ice. *Groundwater* is water stored underground. It is found in *aquifers,* or areas of water under the ground. *Surface water* is stored in oceans, streams, rivers, and lakes. It is also stored in man-made *reservoirs,* which are lakes or ponds created by people. Because there is so much water here, Earth is known as "the blue planet."

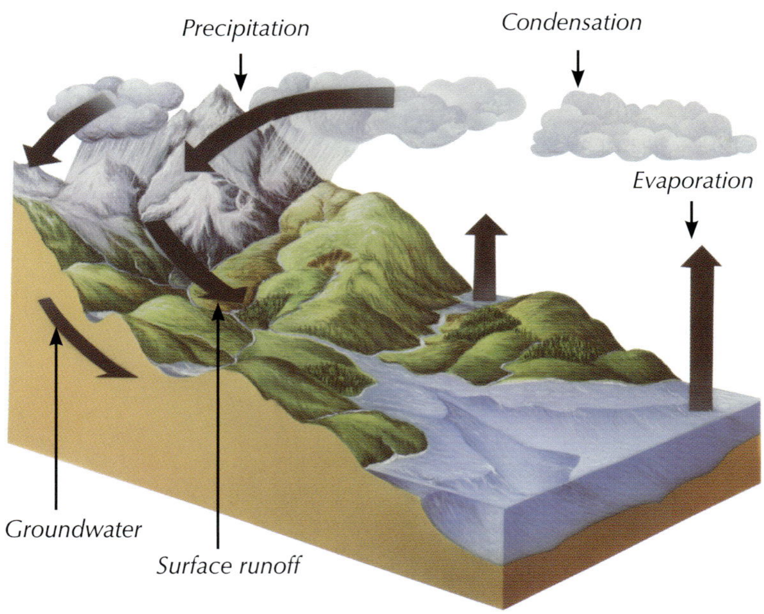

The Water Cycle

Earth has almost the same amount of water now as it had millions of years ago. The water we drink today is the same water the dinosaurs drank. Water on Earth keeps moving through its different states—solid, liquid, and gas. This movement of water on, above, and below the surface of Earth is known as the *water cycle* or *hydrologic cycle*. The water cycle is always happening. It has no starting or ending point. It is made up of several steps:

- **Evaporation:** Heat and energy from the sun warm the water of oceans, lakes, and rivers. The warm water changes from liquid into a gas called *water vapor*. This is known as *evaporation*.
- **Condensation:** In the air, water vapor rises and cools and changes into liquid. The little drops of water in the air form clouds. This is called *condensation*.
- **Precipitation:** When a lot of water vapor condenses into droplets, the clouds become heavy. The air cannot hold the droplets anymore. They fall from the sky. This is called *precipitation*.
- **Collection:** When water falls to Earth, it may fall on oceans, lakes, or rivers. It may also fall on land. On land, water seeps, or soaks, into the ground and becomes *groundwater*.

Oceans

Oceans are large bodies of water that cover about 71% of Earth's surface. Oceans contain about 97% of Earth's water. The five oceans on Earth are the Pacific, Atlantic, Indian, Arctic, and Southern Oceans.

Did you know?

Earth has about 326 million cubic miles of water.

Rivers

Rivers are natural waterways flowing on the Earth's surface. Rivers carry water over land from high points to lower ones. *Altitude* is the height of land above sea level. Mountains are at high altitudes. The sea is at low altitude. Rivers drain their water into the seas and oceans.

Glaciers

Glaciers are thick layers of hard snow and ice that flow like slow-moving rivers. The snow in glaciers builds up over many years. Glaciers are the largest storage of freshwater on Earth. Around 69% of the world's total freshwater is in glaciers and ice caps. If these glaciers and ice caps melt, sea levels could rise by as much as 210 feet. Most glaciers are in Greenland and Antarctica.

Aquifers

Aquifers are underground water storage areas. Aquifers store water between layers of rocks, sand, and gravel. Aquifer water is cleaner than surface water. Rocks and soil filter out (stop from passing through) anything that may be in the water. The water in aquifers can be pumped out by digging wells into the ground. This freshwater can be used in homes and on farms.

Lakes and Reservoirs

Lakes are water bodies surrounded by land. Most lakes are a source of freshwater. However, some lakes are even saltier than the ocean. Reservoirs are man-made lakes. They form when dams are built on rivers.

Usable Water

Usable water is water that we can use in many ways. We use water for drinking, washing, bathing, cooking, and watering plants and food crops. Earth has plenty of water, but not all the water on Earth is usable. Ocean water is salty and therefore not usable. Only 1% of the water on Earth is usable for humans. This usable water can be found in groundwater, lakes, and rivers.

Ways We Use Water

- For agriculture (farming)
- In homes
- By industry (companies)
- To generate thermoelectric power (electricity)
- For mining

What makes water unusable?

- Hazardous (dangerous) waste
- Wastes that have not been correctly handled
- Use of chemical pesticides and fertilizers
- Water that has become mixed with motor oil, grease, and paint

Virtual Water: Embedded Usable Water

Agriculture and industry use the most water. Some water is *embedded*, or stored inside, our food and other products. This is called *virtual water*. For example, about 265 gallons of water is needed to produce 2 pounds of wheat. So the virtual water of this 2 pounds of wheat is 265 gallons. Every day, humans use lots of virtual water. The amount of virtual water used depends on the product.

Did you know?

About 280 cubic miles of water evaporates into the air every day.

Facts You Should Know

- About one gallon of water is used to make a quarter pound of hamburger.
- About 39,090 gallons of water is needed to make a new car, including its tires.
- One inch of rainfall gives about 27,000 gallons of water per acre of land.

Potable Water

Water that is safe and suitable for humans to drink is called *potable* water. Potable water does not contain salt. It is also known as drinking water. Potable water is actually *treated water*. Water from streams, rivers, springs, and lakes is treated to remove impurities. Potable water is not found in the same amounts everywhere in the world. Most of Earth's freshwater is located in North America.

Where does potable water come from?

Potable water comes from two sources: groundwater and surface water. Both of these water sources depend on rain, snow, and other forms of precipitation. Groundwater is an important source of potable water in *arid* (dry or desert) areas. The western states of the United States get their potable water mainly from groundwater reserves.

Who is responsible for potable water quality?

Water can contain around 80 contaminants that can cause health problems. Governments are responsible for providing safe potable water by:

- Checking for contaminants in water
- Making sure there is enough potable water
- Getting potable water to places where people need it
- Recording complaints from people who use their water
- Responding to these complaints

What you can do?

- Know the source of your drinking water.
- Know about the contaminants that pollute drinking water.
- Help with your community water-conservation activities.
- Always watch for chemical spills and leaks. Report these events.
- Follow the directions when using and cleaning up chemicals.

Water Contaminants

Contaminants are impurities that pollute water. They are harmful for humans and for plants and animals. Contaminated water can cause serious diseases. Waterborne diseases are caused by drinking contaminated water. Contaminants can be natural or man-made. Natural contaminants can come from soil or from rocks. Man-made contaminants include chemicals and toxic materials from industrial, agricultural, or *domestic* (household) sources. Some common water contaminants:

- Batteries
- Cleaning solvents (chemicals)
- Disinfectants that are used to kill bacteria
- Explosives
- Grease
- Laboratory chemicals
- Oils (petroleum based)
- Paints, primers, thinners, dyes, and stains
- Photo development chemicals

World Water Day

Potable water is becoming scarce in many parts of the world. About 1.1 billion people do not have access clean water. The United Nations has named March 22 "World Water Day."

Did you know?

At any moment, the air contains about 3,100 cubic miles of water, mainly in the form of water vapor.

How Much Do We Need?

We need water to drink, to produce food crops, and to meet our energy needs. Water is used to water crops, in industries, to produce electricity, for livestock, and for drinking. According to the World Health Organization, each person needs at least five gallons of water every day. People in different areas use different amounts of water. An average American uses about 80–100 gallons of water per day. Water use per person per day in Asia is only 22 gallons, and in Africa just 12 gallons.

Agriculture

We need water to produce our food. About 70% of Earth's freshwater is used for *agriculture,* or farming. Water is used in agriculture to *irrigate* (water) fields. A large amount of water is needed to grow crops.

Industries

Industries also need large amounts of water. Industries that produce metals, wood and paper products, chemicals, and oils use a lot of water. Every year, industries use about 22% of the global usable water.

Domestic Supply

Domestic use of water includes water used in homes every day. Water is used in the home for drinking, cooking, bathing, washing clothes and dishes, flushing toilets, and watering lawns and gardens. Domestic use of water accounts for about 8% of the total water used in a year.

Thermoelectric Power

Thermoelectric power is electricity generated from water. It uses the most water in the United States. Electricity is generated with the help of steam-driven turbines. Thermoelectric power uses 70% salt water and 30% freshwater.

Did you know?

Los Angeles still uses the same amount of water that it did in the 1970s, even though there are one million more people living there.

Livestock

Livestock are animals raised on farms. Beef cattle, dairy cows, chickens, horses, pigs, sheep, and goats are examples of livestock. Livestock produces meat, poultry, eggs, and dairy products. A dairy cow needs about 42 gallons of water per day.

Mining

Mining is the process of taking valuable minerals from the ground. These minerals include coal, oil, gold, silver, copper, diamonds, and other precious stones. Quarrying, milling, crushing, screening, and washing use water. Mining accounts for about 1% of the groundwater used.

Water Use in Various Countries

In Europe, about 54% of water use is by industries. Agriculture uses only 33%. Developing countries use most of their water for agriculture. In Africa and India, agriculture alone uses nearly 90% of all water.

Precious Water: Water Management

Large parts of Earth are covered in water. But most of Earth's water is *saline,* or salty. Salt water is not fit for humans to drink. Drinkable water is a precious commodity. Therefore, it is very important for water to be managed well. Water management aims to limit water use. This is done so that freshwater can be *conserved,* or saved for the future. Water management also helps to determine where freshwater is and how much is available.

World Water Forum

The World Water Forum is a worldwide drive for better water management. This forum is sponsored by the World Water Council. The World Water Forum is held every third year.

The main goals of the forum:

- To highlight the importance of water
- To create awareness about water issues
- To propose possible solutions
- To get governments to act where needed

Water Management Policy

The world's population is growing. This will create more water shortages. The scarcity of water is already a problem. To manage water better, governments have made policies to *conserve,* or save, water. **These** *water management policies* aim to:

- Find new areas where usable water is available
- Supply water to water-scarce areas
- Make sure available water is used correctly
- Make people aware of water waste
- Save money and electricity by saving water
- Find solutions for growing water needs
- Maintain quality of surface water and groundwater

Water Management Methods

Water management methods cover three areas: reducing water loss or saving water, increasing awareness among people, and reusing wastewater. Some of these methods:

- Collect rainwater
- Apply xeriscaping (a gardening method) to reduce water waste
- Install new and efficient equipment that treats sewage waste
- Determine how much water is used
- Store and reuse the water released by air conditioners
- Reuse water used for cooling machines in industry

Water Recycling

Water recycling is the reuse of water. Most types of wastewater can be recycled. Some wastewater can be recycled directly and reused. Other types of wastewater can be recycled by water treatment plants. Treated wastewater is not drinking water. It is used to irrigate fields, flush toilets, and in industries and landscape irrigation.

Benefits of Water Management

- Saves precious water
- Preserves water quality
- Saves wildlife
- Reduces floods
- Helps *aquatic* animals (animals that live in water)

Did you know?

If all the water on Earth were poured onto the United States, it would form a 90-mile-deep layer of water.

Water Shortage: Population Growth

On October 12, 1999, there were 6 billion people living on Earth. Each year, 80 million or more people are added to the world's population. The increasing population needs more freshwater.

Factors Affecting Water Shortage

- Population growth
- Climate change
- Food production
- Water quality
- Immigration
- Poverty (not having enough money or resources)
- Urbanization (more people living in cities)

Immigration and Water Supply

Immigration is moving from the land where one was born and settling in a new country. Immigration increases the population of a country. This can create water shortages.

Did you know?

By 2025, about 48 countries, or 35% of the global population, will face water shortage.

Urbanization

Urbanization is the movement of people to towns and cities. People move to *urban*, or city, areas in search of new opportunities such as jobs. Urbanization leads to more people in a smaller area. Cities may become overpopulated. Water is scarce in these cities.

Climate Change

Climate change is the change in an area's climate due to global warming. Earth is getting warmer. This causes more droughts in some areas and heavier precipitation in others. Increasing temperatures also cause glaciers to melt. All of these factors affect water availability.

Population Growth in Asia

The population of India, China, and Pakistan will increase by about a billion people by 2050. By 2050, India is expected to add 519 million people. China will add 211 million people. And Pakistan will add nearly 200 million.

Water Shortage: Drought

Drought is a natural event. It is usually a long period of dry weather. This causes a serious shortage of water. Drought occurs due to lack of rainfall. It harms the plants, animals, and people of the area. During droughts, there is very little moisture in the air. This causes breathlessness, dryness, and thirst. Drought may last a few months or several years. Droughts affect food production, economic development, and the environment.

Deadly Droughts

The effects of drought can be even worse than other natural disasters. The 1984–85 droughts in the horn of Africa led to a *famine* because the crops did not grow. The famine killed about 750,000 people.

Causes of Drought

- Different patterns of rain or snow in an area
- Water supplies are too low to meet needs
- Change in wind patterns
- Change of ocean currents

Effects of Drought

- Soil *erosion,* or washing away of soil into rivers
- Invasion of destructive insects and other pests
- Rare animals and plants become *endangered* (at risk of dying out completely)
- Wildfires
- Dust storms
- Loss of crops and livestock
- Movement of humans, animals, and birds away from an area
- Loss of swampland, rivers, and lakes

Three Types of Drought

1. *Meteorological drought* occurs due to low rainfall.
2. *Hydrological drought* occurs due to shrinking water levels.
3. *Agricultural drought* occurs due to less moisture in air and soil.

Meteorological Drought

Meteorological drought occurs when rainfall decreases over a period of time. Less rainfall leads to less soil moisture. This affects vegetation. The definition of what makes a drought is very different for different places. This is because some areas are drier than others.

Hydrological Drought

Hydrological drought is the result of decreased rainfall and its effects on water levels. As rainfall decreases, the water levels in rivers, reservoirs, lakes, and aquifers also decrease. Usually hydrological droughts occur after meteorological droughts. They reduce hydroelectrical power production and other water uses.

Agricultural Drought

Agricultural drought is the result of decreased rainfall. Dry air and less moisture in soil affect crops a great deal. Lower water levels means less water is available to irrigate crops. Wasting water is also an important reason for agricultural droughts. Usually these occur after meteorological and hydrological droughts.

Did you know?

Nine of the world's countries possess 60% of Earth's available freshwater supply. They are Brazil, Russia, China, Canada, Indonesia, the United States, India, Colombia, and the Democratic Republic of Congo.

Water Shortage: Water Prices Increasing

Water is now a precious commodity. The short supply increased demand for water also increase water prices . In recent years, more water is used in homes. Water utility companies have raised their prices. The price of clean water will likely continue to climb.

Determining Water Prices

The price people pay for water comes from three main factors: (1) the cost of transporting the water from its source to the people who use it, (2) demand for the water, and (3) price subsidies. *Subsidies* are amounts that governments pay to keep the price of something, such as water or food, lower. Treatment of water to remove contaminants also can add to the cost of water.

Reasons for Increased Water Prices

- Growing populations
- Pollution
- Industrial development
- Scarce water
- Climate change
- Poor water management

Population

Water is a limited resource. The population is growing faster than ever before. A growing population means a growing demand for water. As the population increases, the amount of water available to each person is less and less.

Climate Changes

Climate changes happen partly because heat from the sun melts ice. This adds more water to the oceans. The freshwater from the ice becomes saltwater in the ocean. Saltwater cannot be used as drinking water.

Water Demand

Everyone on Earth needs water to live. In some drought areas, people do not have enough money for underground water pipes. Wells are their only source of water. As the demand increases, water shortages occur.

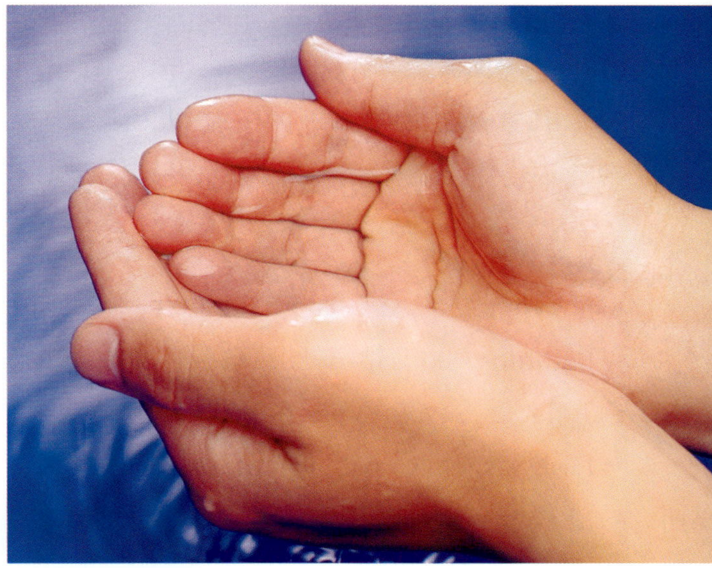

Facts About the Global Water Crisis

- Water-related diseases are one of the leading causes of sickness and death in the world.
- About 88% of all diseases are caused by unsafe drinking water, poor sanitation, and poor hygiene.
- Nearly 50% of the water supply projects in the developing world fail.

Did you know?

In the last five years, water prices have increased by about 32% in the United Kingdom, 45% in Australia, 50% in South Africa, and 58% in Canada.

Water Wasters

People waste water both indoors and outdoors. Wasting water resources harms the Earth. Water is the most important resource. It is needed for most things we do. We waste water every day while cleaning, cooking, and drinking. Wasting water also results in rising costs for other things, such as electricity. By saving water, we can also save money.

Water Wasters

- Showers (2–5 gallons per minute)
- Bathroom sinks (2–5 gallons per minute)
- Toilet (2–5 gallons per flush)
- Dishwasher (25 gallons per load)
- Kitchen sink (2–5 gallons per minute)
- Washing machine (30 gallons per load)

Water Wasted in One Month from Leaks

Source	Gallons Wasted Per Month
A slow steady drip	350 gallons
A fast drip	about 600 gallons
A small stream	2,000–2,700 gallons
A large stream	4,600 gallons

Water Use

Indoor water use in the average single-family home is 69.3 gallons per day in the United States.

Use	Gallons per Person	Percentage of Total Daily Use
Showers	11.6	16.8%
Washing machines	15.0	21.7%
Dishwashers	1.0	1.4%
Toilets	18.5	26.7%
Baths	1.2	1.7%
Leaks	9.5	13.7%
Faucets	10.9	15.7%
Other domestic uses	1.6	2.3%

Putting in efficient water systems and checking for leaks can reduce water use in homes by about 35%. This reduces use to about 45.2 gallons every day.

Use	Gallons per Person	Percentage of Total Daily Use
Showers	8.8	19.5%
Washing machines	10.0	22.1%
Dishwashers	0.7	1.5%
Toilets	8.2	18.0%
Baths	1.2	2.7%
Leaks	4.0	8.8%
Faucets	10.8	23.9%
Other domestic uses	1.6	3.5%

Did you know?

Since 1960, use of water for irrigation has increased by more than 60%.

Water-Saving Etiquette

Saving water is important. People need water for many different uses, from drinking to cleaning. Most of the water used in homes in the United States is used for flushing toilets and bathing. Laundry, cleaning dishes, cooking, and drinking also use a lot of water.

Ways to Reduce Outdoor Water Waste

- Do not overwater lawns and gardens.
- Water lawns and gardens early in the morning.
- Use a layer of mulch around plants.
- Regularly check for leaks and fix them.
- Use a timer on sprinklers.

Ways to Reduce Indoor Water Waste

- Regularly check for leaks.
- Do not run water when not needed.
- Install an energy-efficient hot-water heater.
- Conserve water and reuse when possible.
- When cleaning dishes by hand, fill a bowl or sink instead of letting water run.

Conserving Water When Washing Vehicles

- Wash the vehicle on grass instead of on pavement or sidewalks.
- Give the vehicle a sponge bath.
- Use a bucket of soapy water for washing the vehicle instead of running water.
- Clean your vehicle with a hose that has an automatic shutoff nozzle.

Recycle Water

Most of the water used in homes can be recycled. Toilet water cannot be used again. Other wastewater can be. The main sources of wastewater collection in a home:

- Washing machine
- Sink
- Bathtub
- Dishwasher
- Shower

Did you know?

An average person wastes about 30 gallons of water per day without knowing it.

Water-Saving Ideas

- Take shorter showers.
- Choose plants that need less water.
- Recycle water.
- Do not run water when it is not needed.
- Use a broom instead of a hose for cleaning sidewalks and driveways.
- Start a compost pile instead of putting food scraps in the garbage disposal.
- Do not buy toys, such as squirt guns, that use water.

Saving Water in the Bathroom

The bathroom usually has a bathtub, shower, sink, and toilet. Most indoor water use occurs in the bathroom. Toilet flushing uses 41% and bathing uses 33%. In the United States, water used in baths, showers, toilets, and sinks accounts for about 75% of the total indoor water use.

Why is it important to save water in the bathroom?

Wasting water means a water crisis is likely in the future. We should save water for the future of our planet. By saving water, we can also save on our energy bills. There are many ways to save water in the bathroom. For example, putting in low-flow showerheads turning off the water while brushing teeth can save a lot of water.

Dos and Don'ts

- Install water-saving devices.
- Fill the tub while taking a bath.
- Run water only when needed.
- Install an ultra-low-flush toilet.
- Don't flush the toilet unless it is needed.
- Take showers instead of baths.

Good Water-Conserving Habits

- Recycle and save water.
- Always check for leaks in toilets.
- Insulate hot-water pipes, usually by wrapping them with material that helps hold the heat in.
- Take shorter showers.
- Turn off water while brushing teeth and shaving.
- Bathe small children together.

Did you know?

If every American reduced their water use by one gallon of water every day, we would conserve the same amount of water taken every two days from the Great Lakes.

Saving Water in the Kitchen

Kitchens contain cooking devices such as stoves and microwaves. There is also washing equipment and a sink. Large amounts of water are used in the kitchen. Activities in the kitchen use about 15% of water used in the home.

Saving Water While Cooking

- Use less water for cooking.
- Always cover pans or pots while cooking.
- Use leftover cooking water for soups or sauces.
- Defrost frozen vegetables and fruits in the refrigerator. Don't wash them under a running tap.

Saving Water While Washing Dishes

- Run the dishwasher only when it is fully loaded.
- When washing dishes by hand, fill the sink or a bowl instead of using running water.
- Soak pans or bowls instead of scrubbing them under running water.

Good Water-Conserving Habits

- Install flow-controlled aerators (mixes water with air) on taps.
- Do not use the sink for garbage disposal. Use the garbage can.
- Wash dishes by hand instead of using a dishwasher.
- Always turn taps off tightly.
- Check for leaks and fix them.
- Use fewer dishes, pots, and silverware.

Did you know?

About 25 gallons of water can be saved by using a filled sink or bowl rather than running water from a faucet.

Saving Water in the Garden

People grow flowers, fruits, vegetables, and other plants in gardens. On average, 50% to 70% of home water is used outdoors for watering lawns and gardens. Saving water in the garden saves money by reducing water bills.

Why is it important to save water in the garden?

Saving water in the garden will save about 750 to 1,500 gallons of water every month. It can also keep the garden healthy. Watering plants early in the morning helps stop the growth of fungus and reduces evaporation. Using a layer of mulch around plants also helps reduce evaporation.

Good Water-Conservation Habits

- Install a drip irrigation system. This waters plants better and uses less water.
- Group plants according to their watering needs.
- Water the garden only when needed.
- Put a water-recycling system in your garden.

Dos and Don'ts

- Water the roots and soil rather than spraying water on the leaves and stems.
- Regularly check garden taps and pipes for leaks. Fix them quickly.
- Do not water the garden on windy days. Wind will blow the water to where it is not needed.
- Choose plants that normally grow in your area (these are called *native plants*).
- Do not overwater the garden.
- Use a watering can rather than a hose.

Use Graywater in the Garden

Graywater is water that has been used in the home. This water includes wastewater from sinks, showers, laundry, and the kitchen. Graywater does not include water from toilets. It often contains soaps and fats. Graywater can be used in gardens during droughts or when there is a shortage of water. Graywater can be used in the garden in the following ways:

- Add graywater directly to the soil instead of using a sprinkler.
- Do not use graywater for seedlings or young plants.
- Use graywater on larger areas. Also spray freshwater, which helps keep salts from increasing in the soil from the graywater.
- Water only the stronger plants with graywater.
- Use graywater only when it is cool.
- Do not apply graywater directly to leaves or stems; use it on roots and the surrounding soil.
- Plants that are not eaten can be watered with graywater from baths and showers.

Did you know?

Watering the garden with a sprinkler wastes a lot of water. A sprinkler running for two hours can use about 500 gallons of water.

Benefits of Saving Water in the Garden

- Saves money by reducing water bills
- Protects the environment
- Healthy and beautiful garden

Drought-Resistant Plants

Drought-resistant plants do not need as much water and can live even when there is less rain. They can be planted in your garden to save water. Some common drought-resistant plants:

- Eucalyptus
- Pines
- Poppy
- Tulip
- Juniper
- Lavender
- Marigolds

Saving Water on the Lawn

Keeping a green lawn takes a lot of water. About 32% of water used outdoors at homes is used in caring for lawns. The right amount of watering helps to build a healthier lawn and saves water.

Lawns Help the Environment

Lawns are good for the environment and for human health. They act as filters, trapping contaminants in rainwater before it goes through the soil into groundwater. The microbes (tiny living creatures) in a lawn's roots and soil act as a filter. They trap many kinds of pollutants. Lawns also have an air conditioning effect.

Tips for Saving Water in the Lawn

- Water the lawn only when it is needed.
- Use enough water that it soaks into the soil.
- Place the sprinkler so water can easily reach every area.
- Trim the grass at a higher than normal level. It could help in conserving water.

Dos and Don'ts

- Check for sprinkler or hose leaks and fix them.
- Do not overfertilize the lawn. Use only as much fertilizer as is needed.
- Water lawns in early morning, which helps reduce evaporation.
- Use a timer on the sprinkler.
- Do not water the lawn every day.
- Use mulch around plants, trees, and shrubs.

Did you know?

Placing lawn sprinklers correctly could help save about 500 to 1,500 gallons of water every month.

Xeriscaping

Xeriscaping is a form of landscaping that helps conserve water. The term *xeriscape* comes from two Greek words. *Xeros* means "dry" and *scape* means "a kind of view or scene." It uses landscaping that needs only a small amount of water. It uses different kinds of native and drought-resistant plants. Xeriscaping reduces air, water, and soil pollution. It also saves a lot of money.

The Seven Principles of Xeriscaping

1. **Planning and design:** Planning and design is key to having a good landscape and garden. The plan should show the main parts of the current landscape, including the trees, plants, and shrubs. Plants should be grouped together according to their water and light needs.
2. **Soil improvements:** Soil is the most important factor in xeriscape landscaping. Xeriscape soil should provide plenty of nutrients and water to the plants. Soil can be *improved* by increasing the amount of organic matter. The soil can then absorb and store more water and nutrients. Soil also can be improved by careful use of fertilizers and pesticides.
3. **Efficient irrigation:** *Efficient irrigation* means correct use of water. Not all plants need the same amount of water. Group plants according to their water needs. It is best to water deeply but less often to develop deep roots.
4. **Mulch:** Mulching is needed for xeriscape landscaping. It helps improve how the garden looks. Mulching stops erosion and weed growth. It also helps reduce evaporation. It therefore reduces the amount of water needed. Mulch also decomposes, or breaks down, and feeds the soil.

5. **Maintenance:** Xeriscape landscaping is *low maintenance*. It does not take much work to care for it. Xeriscape landscapes need watering, weeding, pruning, fertilizing, and pest management.
6. **Appropriate plants:** Choosing native plants with low water needs will conserve the most water. Use drought-resistant plants. Their thick, small leaves help them save water.

7. **Turf areas:** Turf grasses are easy to maintain. They prevent erosion and are important for cooling the environment. Turf grasses also provide a play area for children.

Benefits of Xeriscaping

- Water conservation
- Low maintenance
- Little or no use of pesticides
- Reduces disease problems
- Provides a lot of planting options that look nice
- Pollution free
- Raises property values
- Provides places for wildlife to live

Did you know?

Xeriscaping can raise the value of a property by up to 15%.

Saving Water in Public Places

A *public place* is any area or building to which the public has access. Shops, schools, restaurants, hotels, parks, theaters, and clubs are examples of public places. Some European countries, such as Norway, Sweden, and Finland, consider all nature areas public places.

Ways of Saving Water in Hotels and Restaurants

- Put up signs about water conservation in restrooms used by customers.
- Teach employees about water conservation.
- Have an employee keep track of water management.
- Serve drinking water only when customers ask for it.
- Immediately fix dripping faucets and showers to stop leaks.
- Put in flow reducers or adjusting devices in toilets.

Ways of Saving Water in Commercial Buildings

- Install water-saving signs inside office buildings and restrooms used by employees.
- Have an employee keep track how much water is used and saved.
- Install suggestion boxes and seek suggestions from employees and visitors.
- Shut the air conditioning off when it is not needed.

Ways of Saving Water in Schools and Colleges

- Increase awareness among students about the importance of saving water.
- Make water conservation part of what students learn in the classroom.
- Put water-conservation signs in the classrooms and restrooms.
- Check the water supply system often, and fix any leaks.
- Sweep the floors more often than mopping.

Ways of Saving Water in Parks

- Use water for landscaping and gardening only when needed.
- Water in the early mornings or evenings to reduce evaporation.
- Use timers in sprinkler systems.
- Cover swimming pools when not in use to reduce evaporation.

Did you know?

About 95% of the 80 million people added to the world every year come from developing countries. This stresses the world's water supply.

Reclaimed Water and Graywater

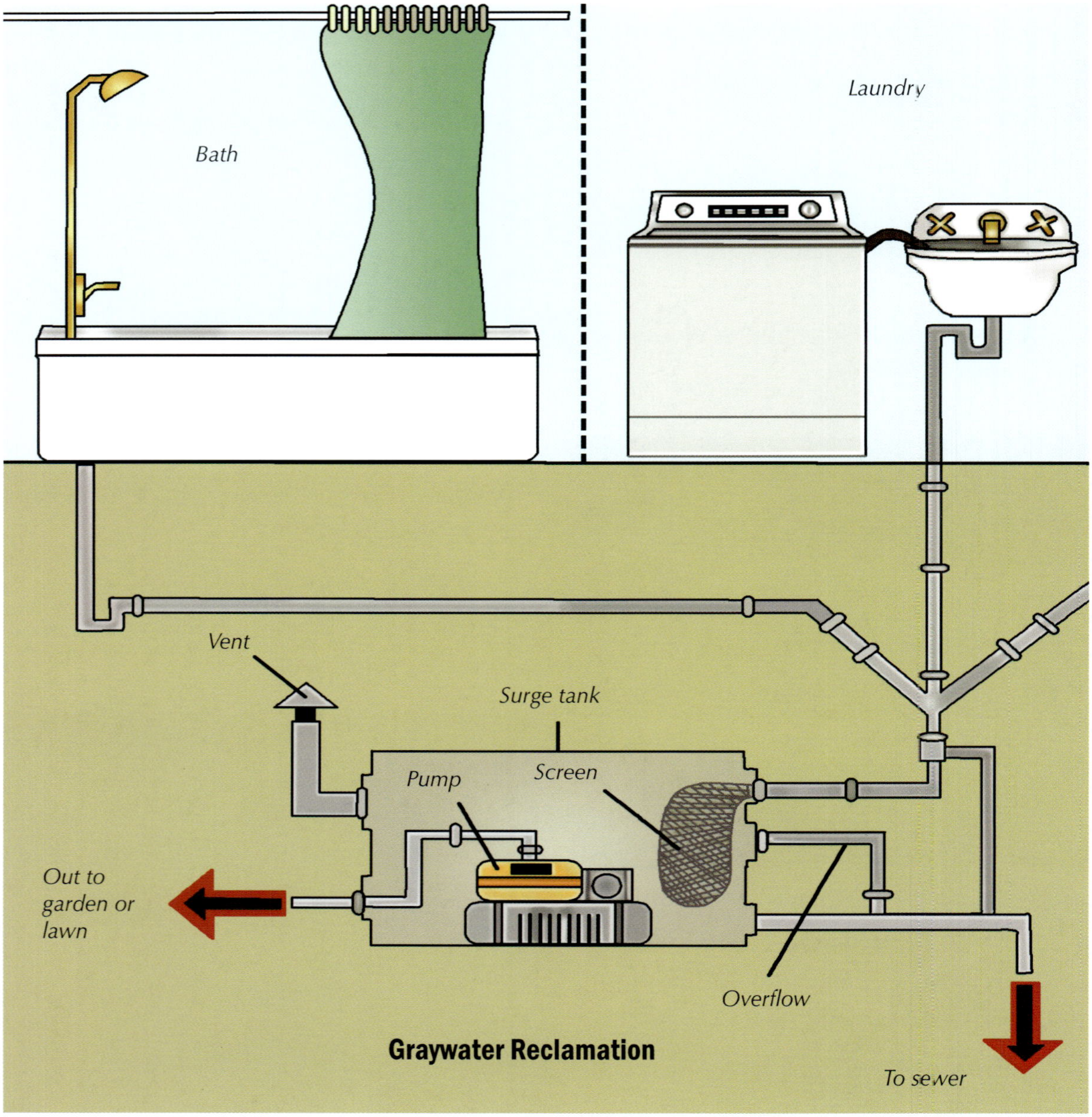

Graywater Reclamation

Wastewater that has been used in homes and that can be reused in other ways is known as *graywater*. It comes from sinks, showers, baths, and the laundry. It does not include water from toilets, kitchen sinks, or dishwashers. Treated graywater is known as *reclaimed water*.

Ways to Use Graywater

- To water plants, lawns, and trees that are not used for food
- To make fountains and ponds
- To sprinkle on dust on public roads
- To water golf courses
- To flush toilets

Dos

- Filter graywater to take out hair and fibers that come from clothes.
- Check your garden often for any damage from organic materials in graywater.
- Use organic shampoos, detergents, and cleaning products that will not make graywater toxic and harmful to plants.
- If graywater has a lot of detergent, it should be mixed with clean water.
- Rotate graywater irrigation with rainwater or potable water to reduce the amount of salts that may build up in the soil.

Don'ts

- Don't drink graywater.
- Don't use graywater where irrigation is done by spraying.
- Don't use graywater to water root or leaf crops such as carrots, beets, and lettuce.
- Do not use graywater if any family members suffer from diseases that are infectious, or pass from person to person. Use of graywater in these cases is not safe and may be against the law.
- Never use graywater from washed clothes spoiled by feces (human waste) or vomit.
- Don't store untreated graywater for more than 24 hours. The bacteria in graywater will make it septic, which is not safe, and could produce a bad smell.
- Don't wash your hands with graywater before eating or drinking.

Did you know?

About 75% of the total wastewater that flows into sewers from homes is graywater.

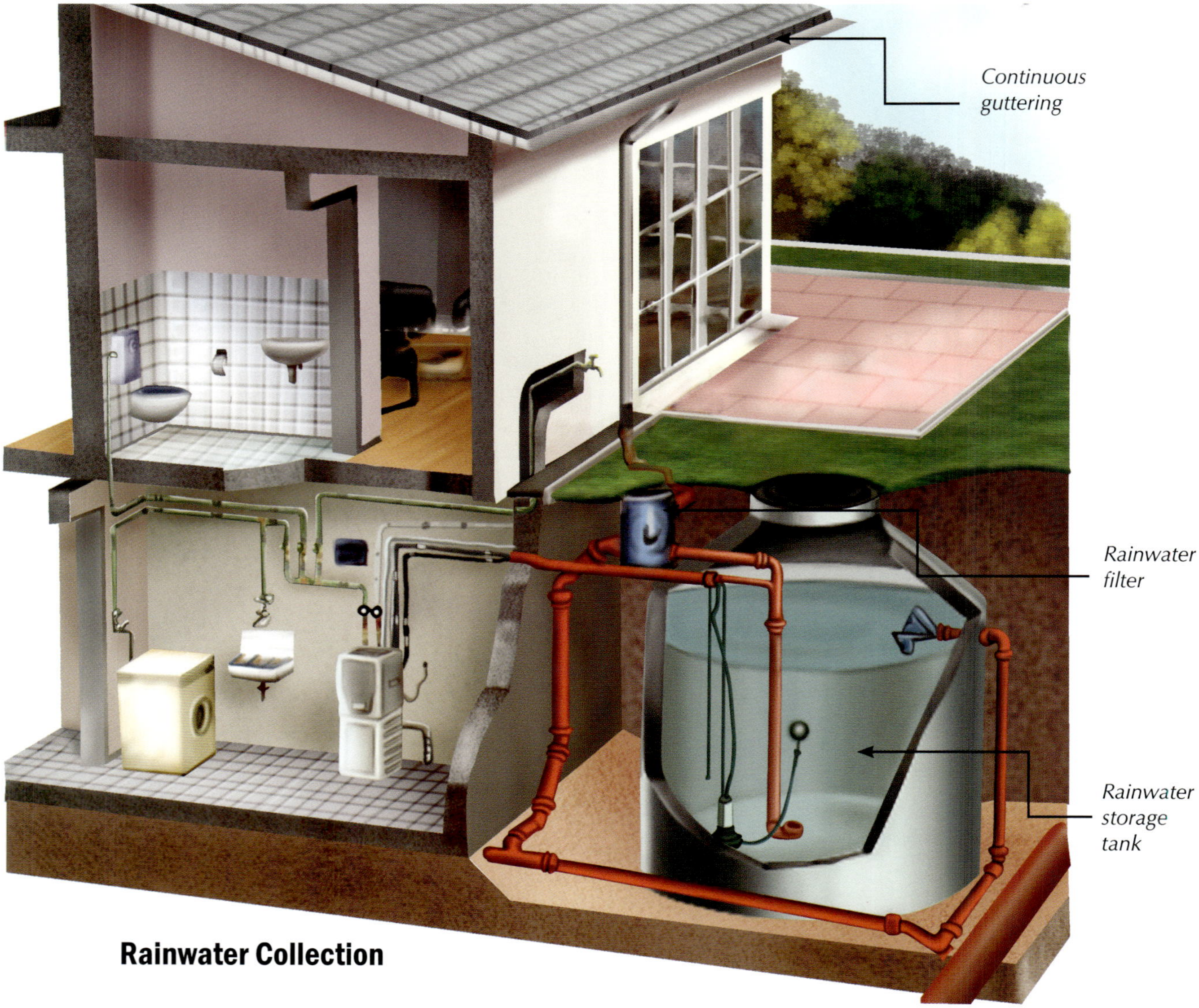

Rainwater Collection

Rainwater Harvesting

Collecting water from rain and storing it for later use is known as *rainwater harvesting*. In this water-conservation method, rainwater is stored in tanks to be used later in homes or on farms. Rainwater is also used to recharge groundwater so that it can be drawn when needed.

Ancient Method

Conserving water through rainwater harvesting is as old as human society. Rainwater was harvested in ancient Palestine and Greece. It was used to water crops. The ancient Romans built their homes in such a way as to catch rainwater.

Uses of Rainwater Harvesting

Harvesting water from rain is useful in many ways. Rainwater can be used for potable as well as nonpotable needs. Potable uses of rainwater include drinking, cooking, bathing, and livestock feeding. Nonpotable uses include filling swimming pools, flushing toilets, washing clothes, and watering gardens and lawns.

Did you know?

About 50% of water used in homes could come from rainwater harvesting.

Advantages of Rainwater Harvesting

- Rainwater is pure.
- It reduces strain on water supplies from other sources.
- Rainwater contains very little salt. It does not cause salt to build up in the soil.
- Rainwater harvesting helps reduce topsoil erosion.
- Rainwater is free of cost and lowers the water bill.
- In cities, rainwater harvesting reduces urban flooding.
- Rainwater harvesting helps improve the groundwater level.
- Rainwater harvesting is most useful where groundwater is scarce or contaminated.

Countries Using Rainwater Harvesting

- Australia
- New Zealand
- Singapore
- India
- Germany
- China

Lakes and Depressions

Lakes are large bodies of water surrounded by land. Lakes are larger and deeper than ponds. Large lakes are often called seas. The world's largest lake is called the Caspian Sea. Most lakes contain freshwater. However, salty lakes are found in dry areas where water evaporates very quickly. The Caspian Sea, the Dead Sea, and Utah's Great Salt Lake are salty lakes.

Where are lakes found?

Most lakes on Earth are found in the Northern Hemisphere. North America has the highest number of lakes of any continent.

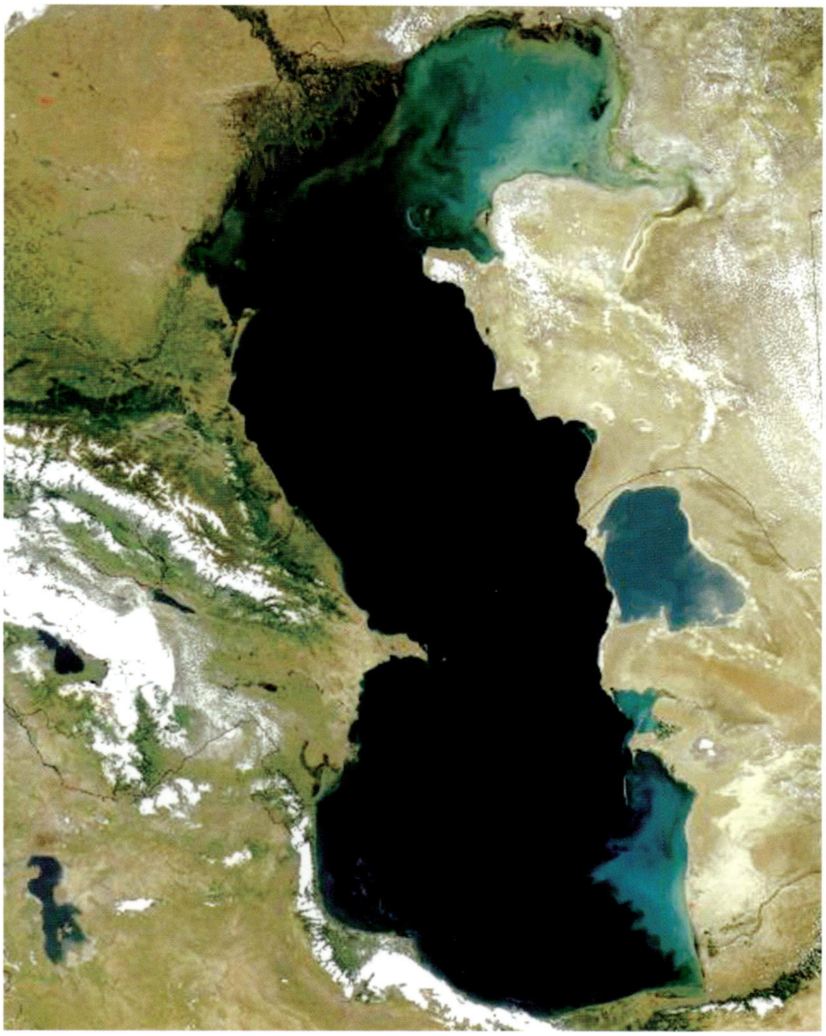

Caspian Sea

Important Lakes

- Caspian Sea, Asia
- Aral Sea, Uzbekistan and Kazakhastan
- Baikal Lake, Russia
- Lake Tanganyika, Africa
- Lake Superior, United States and Canada

Ways of Increasing the Storage Capacity of Lakes

Lakes store a large amount of freshwater. Lakes provide water for electricity generation, irrigation, and navigation. Too much use of lake water shrinks lakes. Sometimes they go dry. To make a lake last longer, its storage capacity needs to be increased. Some ways to improve a lake's water storage:

- Remove silt from the lake during the dry season.
- Create channels so stormwater goes into the lake.
- Stop the inflow of sewage and wastewater from outside areas.
- Try not to take baths, wash clothes, or clean vehicles near lakes.
- Plant trees near lakes to stop soil erosion.

Ways of Improving Water Quality

Lakes are important sources of drinking water. For safe drinking water, the quality of water has to be improved. Fish control weeds in the water by feeding on them. By controlling weeds, fish help to keep the water clean and make it fit for drinking.

Did you know?

Earth's lakes and streams contain about 52,998 cubic miles of water.

Floodplain Reservoirs

A *floodplain* is an area of land near rivers and streams that floods often. It is naturally flooded during the rainy seasons. A floodplain is formed naturally by the *sediment*, or soil and silt, left by rivers. Floodplains can be used to store water for later use. The stored water in a floodplain reservoir can be used to irrigate fields and to produce hydroelectric power. Through careful use of floodplains, we can conserve a huge amount of runoff water. The floodplains refresh groundwater and filter it every year.

Floodway

A *floodway* is the lowest point in a floodplain. It is also the most dangerous part of a floodplain. This is because it has the deepest and fastest floodwaters. If floodwater is blocked in the floodway, then it may increase the water level. This could become dangerous.

Flood Fringe

The *flood fringe* is the outer part of the floodway. It has standing water instead of flowing water. A flood fringe has slow water flow. The water is not as deep as in a floodway.

Yamuna Floodplain

The *Yamuna floodplain* is located in Delhi, India. The population of Delhi is growing. This increases the need for freshwater. The Yamuna floodplain stores monsoon water. It is used later for most of the drinking water needs of Delhi. Floodplains provide a good way to supply clean drinking water.

Floodplain Benefits

- Provides good farmland
- Recreational resource (place to enjoy activities such as swimming)
- Provides a place for wildlife and plants to live

Did you know?

During the 20th century, our water use has increased sixfold, which is twice the rate of the world's population growth.

Flora and Fauna of Floodplains

Plants
- Cattails
- Tamarack trees
- Water lilies

Amphibians
- Frogs
- Salamanders

Birds
- Canadian geese
- Cardinals
- Cranes
- Ducks
- Great blue herons
- Osprey
- Egrets
- Owls
- Red-tailed hawks
- Trumpeter swans
- Blackbirds

Insects
- Dragonflies
- Mosquitoes
- Whirligig beetles
- Waterboatmen
- Monarch butterflies

Mammals
- Bats
- Beavers
- Raccoons
- River otters
- Mink

Reptiles
- Turtles

Quarry Reservoirs

Large, natural water-storage areas known as *quarry reservoirs* are often formed in rock quarries. A *quarry* is a shallow, open pit left after rocks or minerals are removed. Quarry reservoirs are deep and can store lots of water. They store water that overflows from water channels and rivers. They decrease the possibility of flooding.

Types of Quarry Reservoirs

- Limestone
- Marble
- Sandstone
- Slate
- Granite
- Shale
- Clay

Benefits of a Quarry Reservoir

- Controls flooding
- Stores water
- Helps the environment
- Helps protect the watershed
- Promotes water conservation
- Improves water quality
- Reduces water pollution

Stone Quarry Reservoir

Stone Quarry Reservoir is located in Carrboro, North Carolina. It can store up to 200 million gallons of water. By 2030, another much larger stone quarry will be ready for use in the area. It can store three billion gallons of water.

Island Quarry Reservoir

The Island Quarry Reservoir is located at Byron Bay, New South Wales, Australia. It was a quarry where blue stone was removed to be used for walkways and patios.

The Thornton Quarry Reservoir

The Thornton Quarry Reservoir is in Illinois. It used to be one of the world's largest limestone quarries. Now it serves as a reservoir to store sewer water during periods of heavy rainfall.

Spring Creek Reservoir

Spring Creek Reservoir is a water storage area in Santa Clara, California. It used to be a stone quarry. Now it is used to control floods. It stores extra floodwater from Springbrook Creek.

Did you know?

At least 50% of the world's wetlands have been destroyed. As a result, more than 20% of the world's freshwater animal and plant species have already become extinct.

Ancient Reservoirs

The oldest man-made structures on Earth for storing water are known as *ancient reservoirs*. These include dams, ponds, lakes, and wells. They have been around since the earliest human societies in Egypt and Mesopotamia.

Unearthing Ancient Reservoirs

- Many ancient reservoirs are buried at a depth of about 90 to 150 feet underground.
- Most ancient reservoirs are located near or linked with rivers. This helps in recharging water by making good incoming channels.

Some Ancient Reservoirs

- The Romans were very good users of water. Around the first century CE, they built many dams near the Mediterranean Sea.
- A *baoli* is a kind of ancient reservoir found in northern India. Baolis are step-wells used to collect and store rainwater. Baolis were also places where people gathered at that time.
- Most of the ancient temples in southern India have large tanks on their lands. These tanks are fed either by rainwater or by underground springs. They supplied drinking water and also water for use in religious ceremonies.

Advantages of Ancient Reservoirs

- Ancient reservoirs have sand and gravel layers that increase the amount of water they can store.
- Ancient reservoirs are a fairly good source of recharging groundwater. They often serve as flood channels or aqueducts.
- Ancient reservoirs clean water well, as their surface is usually made up of sandy soil.

Did you know?

The earliest ancient reservoirs were dams. Marib Dam in Yemen (built more than 2,700 years ago) was built by the Mesopotamians.

Paleo-Channels

Old abandoned river courses are known as *paleo-channels*. *Paleo* is a Greek term meaning "ancient or something old." A long time ago, paleo-channels were active rivers or streams. Natural or man-made disasters made them dry up. Some also became dry due to changes in river flow. Paleo-channels have good underground water storage.

Paleo-channels as Places to Recharge

Paleo-channels are good for water storage. People in different places can use them during floods or for rainwater harvesting. Most paleo-channels are linked with rivers, creeks, or channels that still have flowing water.

Paleo-channels as Aquifers

Due to natural disasters and climate changes, paleo-channels were long ago buried or shifted to other places. They are a rich source of aquifers. With the help of satellites, geologists try to find paleo-channels. During floods or heavy rainfall, water can be turned toward these abandoned channels. This helps conserve water.

Types of Paleo-channels

Paleo-channels can be divided into four categories:

- Upland paleo-channel belts
- Strip-shaped upland paleo-channels
- Trough-shaped depression paleo-channels
- Ancient riverbeds

Advantages of Paleo-channels

- They provide information to geologists about different places available to store water.
- They provide information about past times.
- They can help geologists predict future climate changes.
- In coastal areas, paleo-channels can provide a connection between freshwater aquifers and the sea. This makes it easier to bring saltwater inland or take freshwater out to sea.
- They often contain deposits of minerals.

Did you know?

The world has about 45,000 large dams that irrigate crop fields producing almost 16% of the world's food.

Check Dams

Small dams called *check dams* are sometimes built on small rivers and streams for harvesting water. Check dams collect rainwater during monsoons or other heavy rains. They help to refill the groundwater level. They store water for future use during the dry season, for irrigation, for livestock, and for use in homes. Check dams can be built at a low cost.

Types of Check Dams

Rock Check Dams

Rock check dams are small dams built across a drainage path. They usually are made from large rocks. Rock check dams can control soil erosion. They are stable and rarely slip or slide.

Log Check Dams

Log check dams are dams built from logs. These dams are used only for short times. Log check dams are built within a channel or ditch or across a drainage path. A log check dam is used to stop soil erosion. They also help slow the speed of flowing water. This allows sediment in the water to settle to the ground.

Did you know?

The United States has more than 2.5 million dams.

Sandbag Check Dams

Sandbag check dams are small check dams that are meant to be used for a very short time. The sandbags are filled with sand and laid across a drainage path, ditch, or channel. These check dams are easy to build.

Maintenance of Check Dams

Check dams should be inspected regularly after every rainfall. If there is any damage, the dam should be fixed immediately. Sediment that has built up behind the dam should also be removed when needed.

Advantages of Check Dams

- Do not cost much , easy to build
- Easy to build
- Stop soil erosion
- Low maintenance
- Increase soil moisture and vegetation
- Reduce damage from flash floods
- Easy availability of drinking water

Disadvantages of Check Dams

- Building and removing check dams can damage vegetation
- Used only in small open channels
- Need high maintenance during high-speed flows
- Not used in live channels or streams

Village Ponds

Man-made water reservoirs known as *village ponds* are the main source of drinking water for people. Village ponds are also used for bathing, washing clothes, and providing drinking water for cattle. Village ponds are found in most parts of the world.

How to Increase Storage Capacity of Ponds

- Deepen the pond by removing silt with the help of dragline machines.
- Do not dump garbage in or near the pond.
- Plant grasses along the sloping edge of the pond to reduce soil erosion.
- Make channels to let stormwater run into the village pond.
- Ponds can be connected with rivers and deep trenches to increase storage during flood season.
- Remove extra vegetation, roots, and organic matter from the pond.

Grasses planted along the perimeter of a pond

Duckweed

Improving Water Quality of Ponds

- Keep sewage away from the village pond.
- Control fishing. Fish clean the pond.
- Introduce aquatic plants such as duckweed. These plants can reduce pollution and provide food for fish.
- Cover the edges of the pond with topsoil and plant quick-growing grasses. Do not plant trees on the edges of the pond.

Did you know?

Most people in developing countries walk at least three hours to get water.

Ecoparks

Manmade wetlands known as *ecoparks* are built to treat water. They are also known as *artificial wetlands*. Ecoparks are similar to natural wetlands but within a more controlled environment. In ecoparks, water, wildlife, microorganisms, and the environment work together to treat wastewater. These wetlands are a good way to remove pollutants from municipal (city and town) and industrial wastewater.

What are wetlands?

Wetlands are areas that are covered by water, either during some seasons each year or all the time. They occur where the land is covered by shallow water. Wetlands are also known as *swamps, marshes, bogs, fens, wet meadows,* and *sloughs*. Wetlands are an important kind of ecosystem found throughout the world. They are home to many kinds of plants and animals not found elsewhere.

Why make artificial wetlands?

Artificial wetlands are created to treat or improve the quality of wastewater. They remove many pollutants from contaminated water. Artificial wetlands are easy to build. They cost less than conventional sewage treatment plants. Artificial wetlands allow reuse and recycling of water. Artificial wetlands also attract wildlife.

Types of Artificial Wetlands

Subsurface Flow Systems

Subsurface flow systems are designed with underground water flowing through sand or gravel. They have a sealed basin with a lower layer of rock or gravel. Root systems go through to the bottom of the bed. These are also known as *root-zone systems, rock-reed filters,* or *vegetated submerged beds*. They can stop bad smells and other problems.

Surface Flow Systems

Surface flow systems are designed with water flowing above ground. Surface flow systems have a shallow basin with soil or another material to support the roots. They are often known as *free-water-surface wetlands.*

Vegetation

Artificial wetlands have both *vascular* and *nonvascular* plants. Vascular plants help treat wastewater. They absorb carbon and other materials. Artificial wetlands are also planted with non-woody plants. These include bulrushes, cattails, reeds, and a number of broad-leaved species.

Did you know?

On Earth, freshwater animals are disappearing five times faster than land animals.

Advantages of Artificial Wetlands

- Less expensive than other kinds of sewage treatment
- Reduce bad water smells
- Treat different types of wastewater
- Provide wildlife with places to live
- Increase vegetation
- Reduce contaminants
- Allow water reuse and recycling
- Environment-friendly
- Use less or no energy
- Easy to build

Disadvantages of Artificial Wetlands

- Need larger area
- Attract mosquitoes and other pests
- Unable to treat high concentration, or large amount, of certain pollutants
- Treatment depends on weather conditions

Artificial Wetlands Treat Water from Many Different Sources:

- Agricultural wastewater
- Sewage
- Stormwater
- Water from landfills

Artificial Wetlands Are Different from Natural Wetlands in the Following Ways:

- Do not change in size
- Not connected directly to groundwater
- Hold more sediment
- Plants and other living things develop very fast

Water Contaminants

Things that mix with water and make it unhealthy for drinking are known as *water contaminants*. We call contaminated water *dirty water*. There are different types of water contaminants. They can be naturally occurring organisms or materials, or man-made chemicals. Contaminated water has become a threat to human life.

Microbes

Microbes are living biological contaminants. They are very tiny organisms. Organisms such as bacteria, fungi, and viruses are microbes. Microbes can cause human diseases. Microbes get into water through sewage and animal waste. The following are some water contaminant microbes:

- *Fecal coliforms* and *E. coli* bacteria that live in human and animal fecal waste. They contaminate water. Contact can cause diarrhea, cramps, nausea, and headaches.
- *Turbidity* is the contamination of water due to sediment from soil runoff. Turbid water often contains harmful bacteria and viruses that can cause nausea, cramps, diarrhea, and headaches.
- *Giardia lamblia* is a type of bacteria that contaminates lakes and river water. It causes illnesses such as giardiasis (also known as *backpacker's diarrhea* or *beaver fever*), diarrhea, vomiting, and cramps.

Organic Contaminants

- *Benzene* is an organic liquid used for making paints, plastics, and man-made fabrics. Benzene enters water from factory wastes. It can cause cancer and other illnesses.
- *Alachlor* is an herbicide used to control weeds in crops. Alachlor can harm the liver, kidneys, spleen, and nose and eyelid linings.
- *MTBE (methyl tertiary-butyl ether)* is a chemical added to gasoline to reduce vehicle emissions. MTBE comes into contact with water through motorboats, leaking gasoline tanks, pipelines, and spills. MTBE causes a bad smell and bad taste in water.

Inorganic Contaminants

Inorganic contaminants come from minerals and rocks. These contaminants can get into drinking water from the release of industrial wastes. They may also come from materials used in plumbing and from natural sources.

- *Lead* is a metal that contaminates water from plumbing pipes. Lead slows physical and mental growth in children. In adults, it causes high blood pressure and harms the kidneys.
- *Asbestos* is a material used for roofing, brake pads, and many other things. Asbestos in drinking water can cause lung disease and cancer.
- *Mercury* is a liquid metal used in electrical equipment. Mercury gets into water from industrial waste and soil erosion. Mercury is very harmful to the kidneys and brain.
- *Nitrate* is an inorganic contaminant that gets into water through fertilizer runoff. Nitrate can cause "blue baby syndrome" in infants and shortness of breath.
- *Fluoride* is a mineral added to water to make strong teeth. It also comes from fertilizers and aluminum factories. Too much fluoride in water can cause pain and tender bones.

Radionuclide

Radionuclide water contaminants are radioactive elements that give off rays or *radiation*. A radioactive element decays all the time. Radionuclide contaminants such as radium and uranium occur naturally in soil and rocks. They also can be produced artificially.

Did you know?

Each day, about 500,000 tons of contaminants are released into the rivers and lakes in the United States.

Drinking Water Treatment

The process that makes water clean enough for humans to drink is known as *drinking water treatment*. Most of the water we find around us is *not* safe and clean. Contaminated drinking water can cause diseases such as cholera, typhoid, and dysentery. Drinking water treatment helps to clean water. It also removes bad smells and makes water taste better.

History of Drinking Water Treatment

The idea of drinking water treatment dates back to 4000 BCE. In ancient times, people treated water by boiling it, putting it in sunlight, or straining it. During the 19th century, two scientific discoveries made drinking water treatment popular in Europe. In 1855, Dr. John Snow discovered that people got cholera from drinking contaminated water. In the late 1800s, Louis Pasteur invented the "germ theory." This theory explained that water contains many microscopic organisms that can cause diseases in humans. In 1914, the U.S. Public Health Service set standards for drinking water treatment.

Boiling

Boiling is an easy drinking water treatment method. Water is put into a clean container and brought to a full boil for at least three minutes. Boiled water is normally covered and stored until it is ready to be used.

Advantages

- It is the best method in emergency conditions.
- It kills harmful bacteria and other living organisms that are found in rivers and lakes.

Disadvantages

- It cannot remove harmful elements such as lead, mercury, and asbestos.
- It uses fuel and cooking equipment.
- Boiled water cannot be used immediately. It needs to cool down.
- Particles may remain after boiling.

Distillation

In *distillation*, water is boiled in a container and allowed to evaporate. The water vapor is pure water. It is stored in another container and allowed to cool until it becomes liquid water.

Advantages

- It produces very pure, clean water.
- It can remove nitrates, chloride, and other salts.
- It also kills bacteria in water.

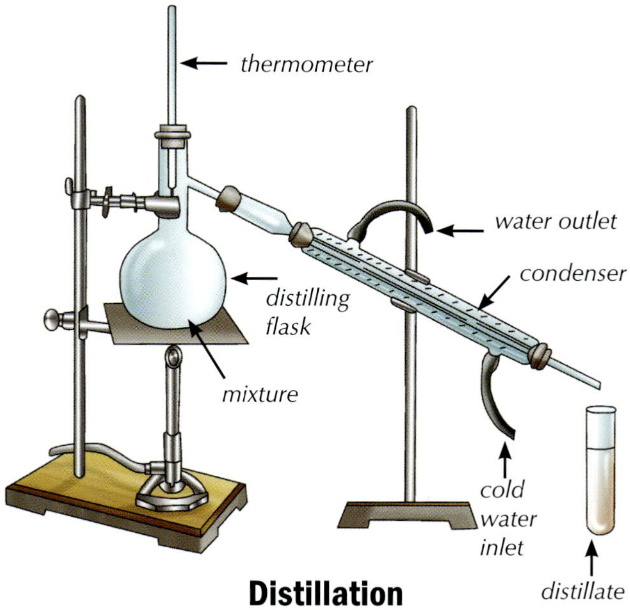

Distillation

Disadvantages

- It is a slow method.
- It uses electricity.
- The devices must be cleaned often.

Filtration

Filtration is a cleaning method that uses filters. Water is passed through a filter that solid materials cannot pass through. It is the earliest drinking water treatment method.

Reverse Osmosis

Reverse osmosis is a very popular drinking water treatment process. In reverse osmosis, water is forced to pass through a *membrane* that has very tiny pores or holes. The membrane is like a filter that allows water to pass through but keeps particles out. It is the best method when water contains high levels of *dissolved* contaminants that are mixed in water very well.

Advantages

- It reduces salt dissolved in water.
- It can remove very tiny organisms such as viruses.
- It is less expensive to operate and maintain.
- It does not use electricity.

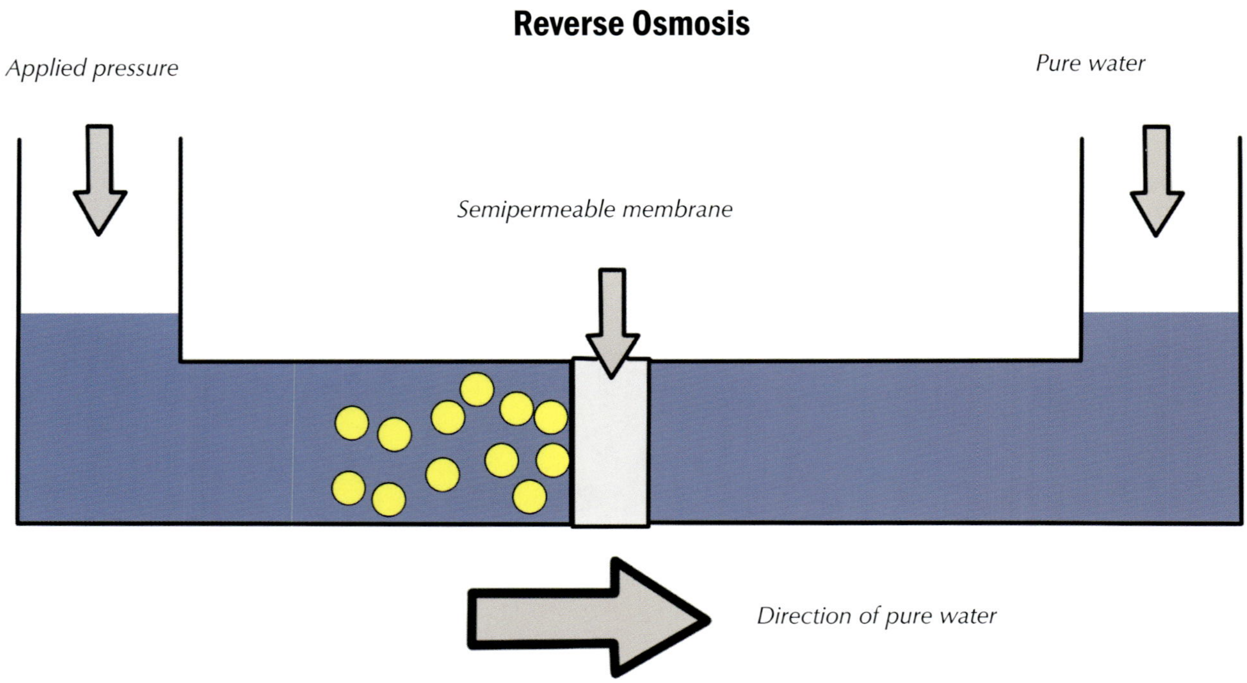

Disadvantages

- It produces wastewater—about two to four gallons of wastewater are produced for each gallon of treated water.
- A damaged membrane cannot be detected easily, but once damaged, it cannot effectively purify water.
- High water pressure is needed for this process. In emergencies, when there is less water pressure, it may not work well.

Ultraviolet Radiation

Drinking water can be treated with *ultraviolet radiation*. In this process, water is exposed to ultraviolet light. Ultraviolet light helps clean water by destroying bacteria and viruses.

Advantages

- It does not leave any taste or smell in the treated water.
- It takes less time.
- It does not affect mineral content in water.

Disadvantages

- It does not work for water contaminated with high levels of solids.
- It cannot remove harmful elements such as lead, asbestos, and chlorine.
- It uses electricity.

Desalination

Desalination is a process of making saltwater fresh and drinkable. Seawater is treated to remove salts and other contaminants. Desalination is a method that includes several other water treatment methods, such as reverse osmosis and distillation.

Did you know?

In the United States, about 48 million people get their drinking water from wells at their homes.

Advantages

- It produces very safe and high-quality drinking water.
- It is useful in areas where there is little freshwater.

Disadvantages

- The process is expensive and uses a lot of energy.
- It produces waste.

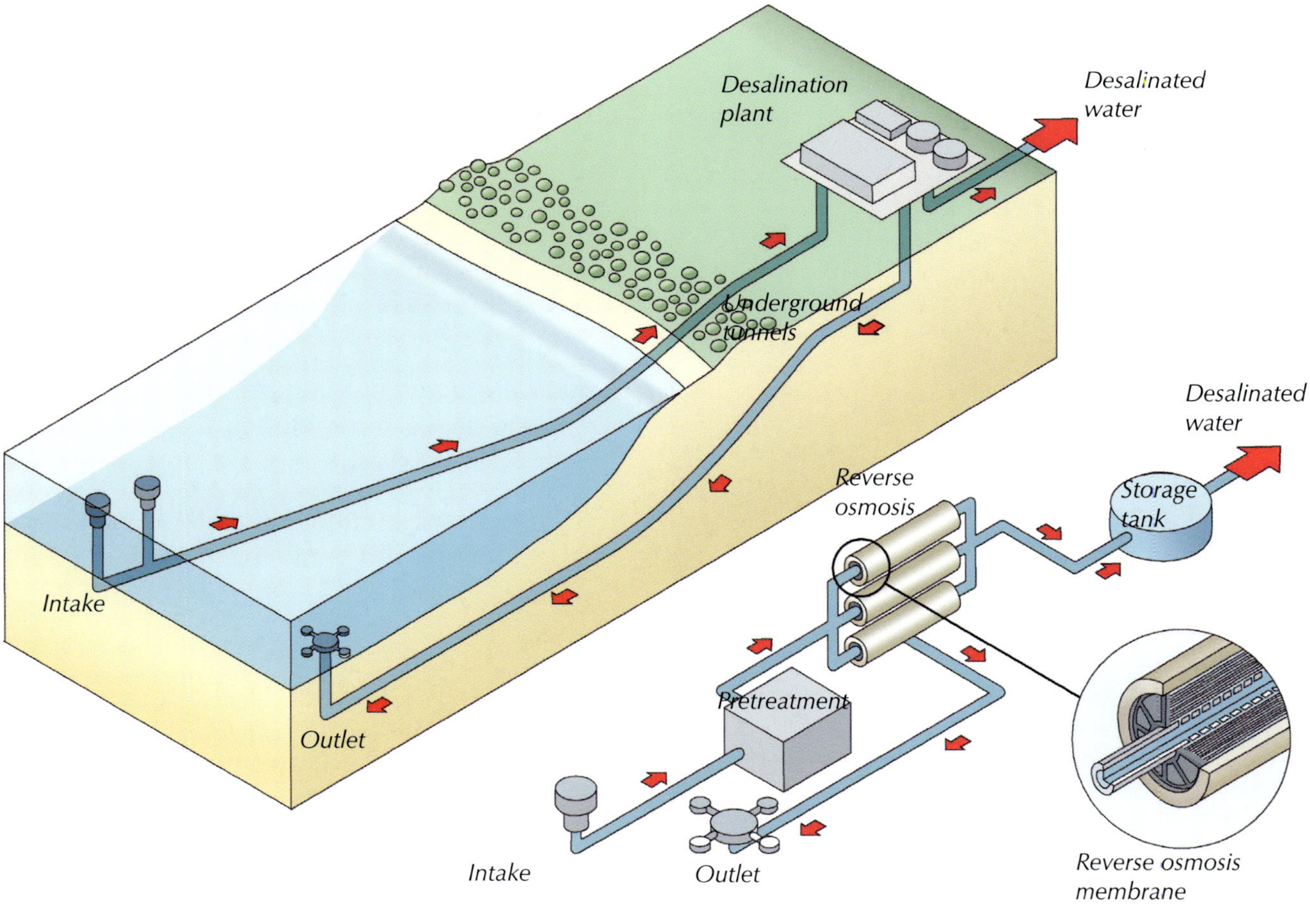

Desalination

Facts and Figures

1. The amount of water on Earth is about the same as it was millions of years ago.
2. The people of the United States drink a total of about 110 million gallons of water per day.
3. About 6.5 gallons of water is used for just one flush of the toilet.
4. An average bath uses 37 gallons of water.
5. Fixing a leaky toilet can save more than 50 gallons of water per day.
6. The average American uses a total of about 140 to 170 gallons of water per day.

7. Showering and bathing use more water than other activities in the home. They use about 27% of the water used in an average home.
8. If every home in the United States had a faucet that dripped once each second, 928 million gallons of water a day would leak away.
9. Approximately one million miles of pipelines and aqueducts carry water in the United States and Canada. That is enough piping to circle Earth 40 times.
10. Some 300 million gallons of water is needed to produce a single day's supply of newsprint for U.S. newspapers.

11. Around the world, about 54% of the available freshwater is used in a year. This figure is expected to go up to 70% by 2025.

12. Water helps to regulate Earth's temperature.

Index

Glossary

agriculture: growing and raising of crops and animals for food

aquatic: ability to that live or grow in or on water

atmosphere: the layer of gases that surround the Earth

bacteria: microscopic, single-celled organisms that can cause diseases

climate: weather in a particular region over a long period

condense: to change from a vapor or gas to a liquid

conservation: protection or management of valuable resources, such as water

contaminate: to pollute by direct contact

decompose: to break down and decay

dissolve: to separate into molecules in a liquid

drought: a long period of little or no rain

ecosystem: a complex community of living things in a physical environment

erosion: wearing away of land or soil by water, wind, animals, and even human activity

evaporation: the process by which a liquid turns into gas

extinct: a species that no longer exists

fertilizer: a chemical substance used to improve the fertility of soil

glacier: a large body of ice that moves very slowly

global warming: an increase in the average temperature of the Earth

groundwater: fresh water found beneath the surface of the ground

habitat: an environment in which a plant or animal normally lives and grows

hemisphere: a half of the Earth on either side of the equator

indigenous: belonging naturally to a given region or ecosystem

irrigation: the addition of water to agricultural land using sprinklers, pumps, or pipes

migrate: move from one place to another to live, spawn or feed

monsoon: a seasonal wind that brings heavy rain and flooding during summer especially in southern Asia

nutrients: substances that help living things be healthy and grow

organism: any living structure—plant, animal, fungus, bacterium—that is capable of growth and reproduction

parasite: an animal that lives on or in another animal that and feeds on it

peat: marshy soil that is made up of decomposed plants

pest: animal or plant that causes harm to other animals or plants

pesticide: any chemical compound that is used to kill pests

pollutant: a substance that contaminates and pollutes air, water, and land

precipitation: falling products of condensation of water vapor in the atmosphere, as in rain, snow, or hail

radiation: rays or waves of energy emitted from the sun

reed: grass-like plants that grow in groups in shallow water

reservoir: a lake that is used to store water

resource: something ready to use

runoff: water that drains or flows off the land into streams and ditches

sediment: small particles such as sand or gravel that settles on land or on ocean floor

topsoil: fertile, upper layer of soil that contains the nutrients needed by plants to grow well

toxin: poisonous substances caused by pollution

treat: to apply special substances to give a particular quality

urban: related to city or town

vapor: the gaseous form of something that is usually liquid

volatile organic compound (VOC): a carbon-rich chemical that evaporates at room temperature

wetland: land such as swamp or marsh